The Truth About Your Food

Be Smart About

# Manufactured Meats

By Rachael Morlock

Cavendish Square

Published in 2023 by Cavendish Square Publishing, LLC
2544 Clinton Street, Buffalo, NY 14224

Copyright © 2023 by Cavendish Square Publishing, LLC

No part of this publication may be reproduced, stored in a retrieval system, or transmitted in any form or by any means—electronic, mechanical, photocopying, recording, or otherwise—without the prior permission of the copyright owner. Request for permission should be addressed to Permissions, Cavendish Square Publishing, 2544 Clinton Street, Buffalo, NY 14224.
Tel (877) 980-4450; fax (877) 980-4454.

Website: cavendishsq.com

This publication represents the opinions and views of the author based on his or her personal experience, knowledge, and research. The information in this book serves as a general guide only. The author and publisher have used their best efforts in preparing this book and disclaim liability rising directly or indirectly from the use and application of this book.

Disclaimer: Portions of this work were originally authored by Stephanie Watson and published as *Mystery Meat: Hot Dogs, Sausages, and Lunch Meats* (Incredibly Disgusting Foods). All new material this edition authored by Rachael Morlock.

All websites were available and accurate when this book was sent to press.

Cataloging-in-Publication Data

Names: Morlock, Rachael.
Title: Be Smart About Manufactured Meats / Rachael Morlock
Description: New York : Cavendish Square, 2023. | Series: The Truth About Your Food| Includes glossary and index.
Identifiers: ISBN 9781502665812 (pbk.) | ISBN 9781502665829 (library bound) | ISBN 9781502665836 (ebook)
Subjects: LCSH: Diet –Juvenile literature. | Food –Juvenile literature. | Meat industry and trade –Juvenile literature.
Classification: LCC TX612.M4 M67 2023 | DDC 641.36–dc23

Editor: Rachael Morlock
Copyeditor: Shannon Harts
Designer: Deanna Paternostro

The photographs in this book are used by permission and through the courtesy of: Cover (main) Brent Hofacker/Shutterstock.com; cover (DNA graphic), back cover, pp. 3, 4, 7, 11, 15, 20, 23, 26, 33, 38-39, 42-43, 44-45, 46-47, 48 Omelchenko/Shutterstock.com; p. 4 N K/Shutterstock.com; p. 5 Postman Photos/Shutterstock.com; p. 6 jreika/Shutterstock.com; p. 9 David Tadevosian/Shutterstock.com; p. 10 silabob/Shutterstock.com; p. 11 MSPhotographic/Shutterstock.com; p. 12 ducu59us/Shutterstock.com; p. 14 stockfour/Shutterstock.com; p. 16 Oleksandra Naumenko/Shutterstock.com; p. 19 Daniel_Dash/Shutterstock.com; p. 20 Cameron Watson/Shutterstock.com; p. 22 Daisy Daisy/Shutterstock.com; p. 25 Aleksandar Karanov/Shutterstock.com; p. 27 Tyler Olson/Shutterstock.com; p. 29 Magic mine/Shutterstock.com; p. 30 Giovanni Cancemi/Shutterstock.com; p. 32 sweet marshmallow/Shutterstock.com; p. 35 medejaja/Shutterstock.com; p. 36 Nataliya Arzamasova/Shutterstock.com; p. 39 Zapp2Photo/Shutterstock.com; p. 40 Amelia Martin/Shutterstock.com.

Some of the images in this book illustrate individuals who are models. The depictions do not imply actual situations or events.

CPSIA compliance information: Batch #CW23CSQ: For further information, contact Cavendish Square Publishing LLC at 1-877-980-4450.

Printed in the United States of America

# Contents

| | |
|---|---|
| Introduction | 4 |
| Chapter One:<br>　More Than Meat | 7 |
| Chapter Two:<br>　Digesting Manufactured Meat | 15 |
| Chapter Three:<br>　Diet and Disease | 23 |
| Chapter Four:<br>　Preserving Your Health | 33 |
| Glossary | 42 |
| Find Out More | 44 |
| Index | 46 |
| About the Author | 48 |

# Introduction

For some people, it's hard to imagine a holiday cookout without hot dogs, a picnic without lunch meat, or a brunch without bacon. Although these foods may be delicious classics, they belong to a group of meats—manufactured meats—that can be harmful when eaten regularly.

Some meat is prepared with methods that make it last longer or taste better. Smoking, salting, **curing**, and **fermenting** are methods that have been used for centuries. Our ancestors preserved meats with these processes to prevent spoiling in the days before refrigeration. Manufactured meats are made in a factory and are also known as processed meats. They include meats from cows, pigs, turkeys, chickens, sheep, goats, fish, and other animals that are raised to produce food. Some common types of processed meats are bacon, hot dogs, jerky, ham, lunch meat, canned meats, meat sauce made in a factory, and sausages.

Hot dogs are a type of sausage that have been a favorite food in the United States since the late 1800s. The country even celebrates National Hot Dog Day every July.

Manufactured meats are familiar foods that can be part of breakfast, lunch, or dinner.

Sausages have been eaten around the world for thousands of years. They are made from ground meat mixed with water, bread, spices, and salt. Factory-made sausages also have **preservatives** and other extra ingredients called additives. Artificial coloring is added before the meat mixture is packed into a casing. Some popular sausages are bratwurst, pepperoni, chorizo, hot dogs, and breakfast sausage.

Lunch meats are another standard kind of manufactured meat. Bologna, salami, and ham are made with red meat before being sliced, packaged, and sold. Turkey and chicken lunch meats usually have less unhealthy fat than those made with red meat. Even though they are leaner, they still contain added ingredients and preservatives.

Eating manufactured meats too frequently can contribute to serious health problems. Some manufactured meats are red meats that are high in unhealthy fats. Most manufactured meats are also generously salted. Diets high in salt and fat can lead to heart disease and diabetes. In addition, processed meats are made with artificial colors and preservatives. The chemicals in processed meats have been shown to cause cancer, and they can lead to serious lung problems. Learning about the factors that make these meats harmful can help you make healthy choices.

Canned meats, such as this ham, are mixed with preservatives and stored in cans that may be lined with harmful substances like Bisphenol A (BPA).

# Chapter One

## More Than Meat

What's in your meat? Meat can be an important source of protein in your diet. It can also bring you much needed vitamins and minerals. If you eat manufactured meat, you may be getting more than you bargained for. Unlike whole, unprocessed meats, such as chicken breast, by the time processed meat arrives on your plate, it's already been through a lot! The meat may have been salted or smoked, or it may have had chemicals and colorings added. Knowing how processed meats are made and what they contain can help you decide if they're foods you want to eat.

### How Sausage Is Made

You may think you're eating pure, whole meat when you bite into a sausage or hot dog, but manufactured meats often contain meat **byproducts**. These byproducts can come from different parts of an animal, such as the heart, kidney, or liver. A hot dog can also contain meat from different kinds of animals combined. Even turkey or chicken hot dogs and other sausages can contain byproducts, such as skin and fat.

Mixed animal body parts can also get into a hot dog when the meat is removed from the animal. Hot dog manufacturers need to separate cow, pig, turkey, or chicken meat from the animal's bones. A lot of the meat put in hot dogs is pulled from the animal using a process called advanced meat recovery (AMR). The meat is scraped, shaved, or pressed to remove it from the animal's bones without breaking them.

Another method for removing meat is called mechanical separation. During this process, the animal's bones with the meat still attached are forced through a device under high pressure. The result is a batter or paste made from meat. Sometimes this meat paste contains tissue from the animal's central nervous system (brain and spinal cord). Because of concerns over mad cow disease—a deadly brain disease in cows that can be passed on to humans through **contaminated** meat—mechanically separated beef is no longer allowed in hot dogs. However, hot dogs can contain up to 20 percent mechanically separated pork and any amount of mechanically separated turkey or chicken. Hot dogs that are made with mechanically separated meat must be labeled to inform consumers that they contain meat processed in this way.

Manufacturers also add **filler** to sausages and hot dogs. Cereal ingredients such as breadcrumbs, oatmeal, or flour help bind the meat together and add weight to the product. Fillers tend to cost less than meat. The more fillers a company adds to its products, the more cheaply it can produce hot dogs and sausages.

## Fat Content

Sausages, hot dogs, and other meats contain fat—sometimes a lot of fat. Manufacturers are allowed to produce sausages that are as much as 50 percent fat, according to the

Sausage casing can be made from the intestines of cows, pigs, or sheep, or made from a protein called collagen found in animal hides. There are also artificial casings that are made from cellulose, a plant-based plastic.

U.S. Food and Drug Administration (FDA). Hot dogs can contain up to 30 percent fat.

Nutrition guidelines advise eating more unsaturated fats than saturated fats. Most nutrition labels and guidelines provide measurements in grams. One hot dog can contain about 13 grams (0.46 ounces) of fat, and 5 grams (0.18 oz) or more of that can be saturated fat. A single slice of salami can have 6 grams (0.22 oz) of fat, and half of that fat is saturated. Some large sausages contain 20 grams (0.71 oz) of fat or more! By eating just one hot dog or sausage, a person can get between 25 to 40 percent of the recommended fat intake for an entire day. Even chicken or turkey hot dogs can contain 8 or 9 grams (0.28 or 0.32 oz) of fat. Low-fat hot dogs are available, but they don't add much, if any, nutrition for the little bit of fat that was removed.

## Extra Salt

According to the American Heart Association, healthy adults should eat no more than 2,300 milligrams (0.08 oz) of sodium each day. Most people get their sodium in the form of salt that

Nutritionists recommend having just one teaspoon of salt a day. That amount includes the table salt you sprinkle on meals and the sodium in any processed foods you eat.

# Food Dyes in Meat

Sodium nitrate is often added to hot dogs to make them look less gray and more appealing. Food dyes made from petroleum can also achieve this effect. Artificial dyes such as Red Dye 40 are regularly added to processed meats like hot dogs to make them look more appealing. Red Dye 40 is also found in cereals, drinks, puddings, gelatin, dairy products, candies, and factory-made baked goods.

The big question is whether artificial dyes are safe for people to eat. Some studies show that artificial dyes can make children **hyperactive**. There is some evidence that removing dyes from the diets of children with **attention-deficit/hyperactivity disorder** (ADHD) can be helpful. Artificial dyes like Red Dye 40 are also known to cause allergic reactions in some people and to contain cancer-causing substances.

Researchers continue to investigate the effects of eating foods with artificial dyes. Meanwhile, some countries have banned the use of certain food dyes. Others, like the United Kingdom, require warning labels on foods that contain them. In the United States, the FDA maintains that artificial dyes are safe to eat in small quantities, so they are found in many processed foods, including meats.

Just as you add food coloring to frosting, artificial dyes are added to processed meat. Although small amounts may be safe, eating many different foods with artificial dyes can have a negative impact on your health.

is added to food. A healthy amount of salt is about 1 teaspoon a day. People who have **high blood pressure** should eat even less salt. Grams, which are part of the metric system of measurement, are often used for measuring salt and other ingredients.

A single hot dog can contain up to 760 milligrams (0.026 oz) of sodium, and a thick slice of ham can contain more than 800 milligrams (0.028 oz). Eating just a few servings of these processed meats every day can put a person well above the daily salt recommendation.

## Additives and Preservatives

Hot dogs, sausages, and lunch meats—such as salami and ham—have a reddish color. However, they didn't start out that way. These meats would look gray when cooked, similar to the color of cooked ground beef.

Because gray food looks less appealing, meat manufacturers add chemicals called sodium nitrate and sodium nitrite to meats. These preservatives not

Even food items marked "nitrate free" may contain celery juice, which is simply a natural source of nitrates.

only add color to the meat, but they also prevent it from spoiling or being contaminated with bacteria that can cause the dangerous disease **botulism**.

Nitrates and nitrites are mixed with meat during the curing process, which is a method of smoking and adding preservatives. Products that are labeled "uncured" do not have added nitrates or nitrites, but they aren't necessarily free from these ingredients. Nitrates and nitrites also occur naturally in some spices and other ingredients used to process uncured hot dogs.

## Food for Thought

1. Have you been taught to think about the fat, salt, and additives a food contains before eating it?
2. Would you be as likely to eat a hot dog that was gray instead of pink, or would you rather eat food that looks better but contains artificial dyes?
3. Do you think food dyes are safe enough to include in food?

Applying salt to fish or meat is a traditional way of preserving food and protecting it from bacteria. However, eating too much salt from processed meat can lead to high blood pressure.

# Chapter Two

## Digesting Manufactured Meat

Healthy foods give you energy and a wide range of nutrients, like vitamins and minerals. In contrast, manufactured meats mostly add fat and sodium to your diet. The ingredients in manufactured meats challenge your body systems and can lead to problems with cholesterol, blood pressure, and your digestive system. Eating manufactured meats every once in a while is generally OK. After all, it's hard to pass up a hot dog at a summer barbecue! However, eating processed meats regularly can have a negative impact on your health over time.

### More About Fat

Hot dogs and sausages can contain 25 percent or more of a person's total fat requirements for an entire day. Adding these foods to a regular diet can increase total fat consumption. Because fat has more than double the calories of carbohydrates or protein, eating additional fat leads to greater weight gain.

Not all fats are created equal. Unsaturated fats, which are found in fish, nuts, and vegetable oils, are liquid at room temperature. This type of fat is considered healthy because it helps rid the body of cholesterol. (Cholesterol is an important part of the body's cells, but having too much of it in the blood can contribute to arterial disease.)

Saturated fats, on the other hand, are solid at room temperature. Examples of sources of saturated fats are cheese,

There are many sources of unsaturated fats for your diet, including fish, nuts, olive oil, and avocados.

butter, and the red meat found in hot dogs and sausages. Saturated fat raises the level of cholesterol in the blood, putting a person at risk for clogged arteries and heart disease.

The most artery-clogging kind of fat is trans fat, which is most often made in a manufacturing plant but can exist in some meats naturally. Companies bubble hydrogen through vegetable oil at very high pressure and very high heat to make the oil thicker and more solid. Products made this way are sometimes described as "partially hydrogenated" on their labels. The process of hydrogenation makes oil more stable so that food can be fried in it over and over again. It also gives foods an appealing taste and texture, and it prevents them from spoiling. Trans fats also raise blood levels of unhealthy cholesterol and lower the levels of healthy cholesterol.

## Balancing Cholesterol

Cholesterol is a waxy, fatlike substance that circulates in the bloodstream. It's used to make cell membranes—the outer layers of cells—as well as some hormones.

However, not all cholesterol is healthy. There is good cholesterol, and there is bad cholesterol. Bad cholesterol is known as low-density lipoprotein (LDL) cholesterol. This type of cholesterol is thick and sticky. It can build up on the inner walls of the arteries. Eventually, the arteries can become so clogged that no blood can pass through them. The bad cholesterol forms a blockage that can cause a heart attack or **stroke**.

Good cholesterol is called high-density lipoprotein (HDL) cholesterol. It acts like an artery cleaner, sweeping extra cholesterol out of the arteries to the liver, where it is removed from the body. This prevents cholesterol from clogging up the arteries and forming a blockage.

Hot dogs and sausages tend to be high in LDL cholesterol and low in HDL cholesterol. Eating too many of them can raise bad cholesterol levels in the body.

## Regulating Blood Pressure

Just as with cholesterol, salt is fine to eat in small amounts. Sodium serves a useful purpose in the body, keeping all of the body's fluids in balance, helping transmit nerve impulses, and making sure that cells, nerves, and muscles all work as they should.

The body has a system in place to ensure that sodium levels stay constant. When there is too little sodium, the kidneys—the body's filtering system—conserve sodium by putting unused sodium back into the bloodstream. Any time sodium levels get too high, the kidneys release the extra sodium into the urine to be removed from the body.

Yet even the hardworking kidneys can get overloaded. When someone eats a lot of high-salt foods, such as hot dogs, potato chips, or lunch meat, sodium starts to build up in the blood.

Sodium attracts and holds water. Extra water ends up in the bloodstream, where it causes the total blood volume to increase. Blood pressure increases as a result. The heart has to work harder to push all of the extra blood through the body. Having high blood pressure can lead to heart disease and strokes.

## Forming Nitrosamines

Sodium nitrite and sodium nitrate are added to manufactured meats during the curing process. In the body, proteins called amines can combine with nitrites and nitrates to form

There's a lot of sodium in lunch meat, like pastrami, and in chips. When people eat a lot of salt, they might feel like their body is more swollen or bloated than usual because the extra sodium holds extra water in the body.

# Meat and the Planet

It's generally true that the foods that are most nourishing for your body are also least harmful to the environment. Eating processed meats and too much red meat in general can have a negative impact on your health. It can also have a negative impact on the environment.

Climate change is worsened by greenhouse gas emissions from animal agriculture, which consists of meat, egg, and dairy production. The production of beef is especially troublesome. The meat industry contributes to climate change by releasing gases like methane, nitrous oxide, and carbon dioxide. Methane is produced by **ruminants** like cows, and fossil fuels are burned during the processing of meat and transportation of animal products.

The United Nations Intergovernmental Panel on Climate Change issued a report called *Climate Change 2022* that examines these issues. It urges people and governments to address climate change by reducing the production and consumption of meat.

Eating less processed meat and red meat is good for you and the planet. The UN encourages people, especially in high-income countries, to eat more plant-based foods. This kind of diet would relieve problems with greenhouse gas emissions and land use while also reducing the risks of cancer, diabetes, and heart disease.

Meat production worldwide accounts for about 60 percent of the world's greenhouse gas emissions from food production. It produces about double the emissions of plant-based foods!

substances known as **nitrosamines**. Usually, only a tiny amount of nitrosamines are formed in the body. How many nitrosamines are formed depends on how the meat is processed, the length of time it is stored, and how long it is cooked, among other things.

Nitrosamines have the ability to cause DNA mutations, or changes to genetic material. Researchers say eating nitrates and nitrites can also increase a person's risk for colon cancer and other cancers.

## Food for Thought

1. Can you think of compromises that would allow you to eat processed meats sometimes while choosing healthier alternatives other times?
2. How many foods can you name that contain unsaturated fats?
3. Moving your body can increase your HDL, or good cholesterol. How do you like to move your body?

Nitrites and nitrates are thought to cause headaches by widening the blood vessels in the head.

# Chapter Three

## Diet and Disease

The high salt and fat content in manufactured meats can result in **obesity**. Almost 20 percent of children ages 2 to 19 are obese in the United States. Adult obesity rates are even higher, at about 42 percent. Obesity and the strain on the body's systems caused by eating too much salt and fat can lead to very serious **chronic** health conditions like heart disease and diabetes. The preservatives and chemical additives in manufactured meats are also connected to increased risk of cancer and the lung disease chronic obstructive pulmonary disease (COPD). The chance of developing these diseases as a result of eating manufactured meats grows slowly over time, but some illnesses can result immediately and directly from eating manufactured meats.

### Obesity

Meat is high in fat. It makes sense, then, that eating a lot of meat can cause a person to gain weight. Gaining too much weight can make a person overweight or obese.

One study found that people who eat a lot of meat are 33 percent more likely to be obese and have extra fat around their middle. Fat in the abdomen is the most dangerous kind of fat because it can increase a person's risk for heart disease, heart attack, and other dangerous diseases. People whose diets are high in processed meat tend to have less room in their diets for healthy, low-fat foods such as fruits, vegetables, and whole grains.

## Diabetes

Weight gain can also increase a person's risk for type 2 diabetes. Being overweight makes the body less sensitive to the effects of the hormone insulin. Normally, insulin helps move sugar from the bloodstream into the cells to be used for energy. When this important hormone isn't working properly, sugar builds up in the bloodstream.

High blood sugar can cause a whole range of health problems, including kidney failure, nerve damage, and blindness. Eating a healthy diet and regularly moving your body can help prevent or even reverse diabetes.

## Heart Disease

Being overweight has yet another side effect. People who are obese are more likely to have heart disease, which can lead to a heart attack or stroke. Obesity boosts a person's heart disease risk by raising levels of unhealthy LDL cholesterol, lowering levels of healthy HDL cholesterol, raising blood pressure, and increasing the odds of developing diabetes, which is itself a risk for heart disease.

A report published in 2020 in *JAMA Internal Medicine* explores the links between processed meat and heart disease. This report is based on a study of almost 30,000 adults over the course of 30 years. It concludes that eating red and processed meat increases the risk of heart disease and early death. A serving of processed red meat in the study was defined as two slices of bacon, two sausage links, or one hot dog. People who ate two servings of these processed meats each week had a 7 percent higher risk of developing heart disease.

To improve heart health, it's best to avoid processed meats and save red meat for special meals, maybe once or twice a month. Practicing a healthy lifestyle with regular movement can make a big difference in preventing heart disease too. It can also be prevented by increasing your intake of healthy, whole foods—fruits, vegetables, whole grains, and beans—in place of meats.

Even unprocessed red meat can lead to heart disease if eaten too often. Red meat is any meat that comes from mammals, instead of from poultry or fish.

## Cancer

Eating meat, including processed meat, increases the risk of developing cancer. Nitrites, nitrates, and other additives in processed meats lead to the production of cancer-causing substances in the body, like nitrosamines. Cooking meat over high heat by grilling, frying, barbecuing, or broiling it can also result in cancer-causing agents.

After reviewing more than 7,000 large studies over 5 years, the American Institute for Cancer Research and the World Cancer Research Fund found that eating too much processed meat and being overweight increase the risk for cancers of the colon, kidney, pancreas, esophagus, uterus, and breast.

## Colorectal Cancer

The more processed meats you eat, the more likely you are to develop colorectal cancer. This type of cancer affects the large intestine of the digestive system. It starts in either the colon, which makes up most of the large intestine and measures about 5 feet (1.5 meters) long, or the rectum, which makes up the last 6 inches (15 centimeters) of the intestines and ends at the anus. The colon is responsible for absorbing water and salt from food matter as it's digested. The rectum stores the remaining waste matter until it can be passed from the body.

Colorectal cancer is caused by changes to the DNA in the cells of the colon or rectum. These changes result in cells that grow unchecked and form polyps, or small growths, in the colon and rectum. Some polyps are harmless, whereas others can grow into dangerous tumors. The risk of developing colorectal cancer is higher for people who have a family history of the disease, are overweight or obese, are not physically active, or have type 2 diabetes. The risk is also higher for people who regularly eat processed meats. Colorectal cancer is currently the third most common cancer in the United States.

One hot dog weighs about 50 grams (1.8 oz). The World Health Organization (WHO) says eating that much processed meat a day is a health risk.

The World Health Organization (WHO) has also determined that processed meat is "carcinogenic to humans," or known to cause cancer. Other known carcinogens include tobacco, alcoholic beverages, and air pollution.

| Carcinogen | Number of Deaths Per Year |
|---|---|
| tobacco | 1,000,000 |
| alcohol | 600,000 |
| air pollution | 200,000 |
| high processed meat | 34,000 |

Shown here are known carcinogens and the number of deaths attributed to them each year, according to the Global Burden of Disease project.

According to the WHO, the strongest link is between processed meat and colorectal cancer. Studies show that eating 50 grams (1.8 oz) of processed meat a day can increase your risk of colorectal cancer by 18 percent. Since the WHO hasn't been able to determine if there is a safe amount of processed meat to eat, some researchers and doctors recommend avoiding it completely or eating it only rarely.

## Lung Disease

Health experts have known for years that smoking cigarettes and other tobacco products can lead to lung diseases. Now they've also discovered that eating sausages, lunch meats, and other cured meats may increase the risk too.

Studies show that people who eat cured meats are more likely to get COPD than people who never eat cured meats, especially if they also smoke and don't often move their bodies. COPD is a disease that is most common in smokers. It causes changes to the lungs that make it more difficult for a person to breathe. Nitrates from cured meats may damage the lungs,

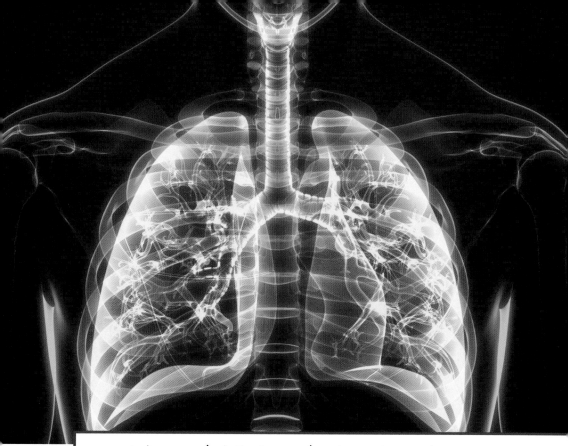

One study revealed that people who eat cured meats 14 times a month are 78 percent more likely to have COPD. This disease affects the lungs, which are shown here.

and people who already have COPD must be especially careful about the meats they eat.

## Bacteria and Parasites

Manufactured meats most commonly have a negative impact on long-term health. Sometimes, though, eating manufactured meats can make you sick right away. Bacteria can contaminate manufactured meat during any stage of production—while it's being prepared, stored, or transported.

*Escherichia coli* (*E. coli*) bacteria live in the intestines of humans and animals, and normally they're harmless. They even help the body make important vitamins. Yet some kinds of *E. coli* produce a poison that can cause symptoms such as diarrhea and life-threatening problems like kidney failure. Hot dogs and other meats that haven't been cooked properly can contain these dangerous bacteria.

*Listeria* is a bacterium that is normally found in soil and water. When animals consume food or water contaminated with *Listeria*, bacteria enter their bodies. People who

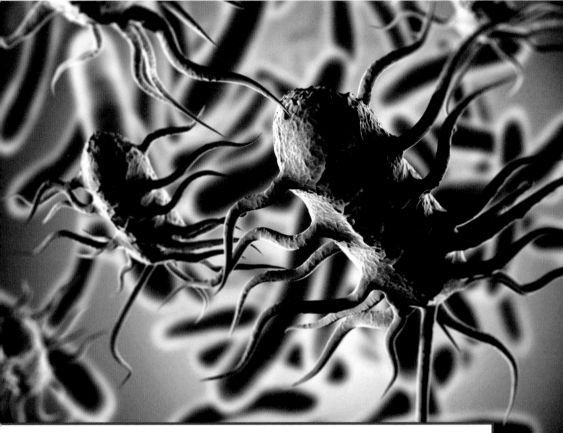

Listeria is a microscopic bacterium that can make you sick. Bacteria aren't always dangerous, though. The body also contains a lot of good bacteria in the stomach and intestines that help with digestion.

eat meat from these animals can become very sick if the meat isn't cooked at a high enough temperature to kill the bacteria. These bacteria can cause an infection called listeriosis.

In most healthy adults, listeriosis causes only mild symptoms, such as nausea, cramps, and diarrhea. In older adults, young children, or anyone with an immune system that doesn't work as well as it should, listeriosis can be very dangerous. Pregnant women who get the disease can lose their baby or give birth too early. It can also cause a baby to be born with very serious health problems.

Parasites—tiny bugs that live off animal hosts—can also live in processed meats and cause illness. *Trichinella spiralis* is a worm found in undercooked pork sausages. People infected with these worms have symptoms such as fever, nausea, stomach pain, and vomiting.

Toxoplasma parasites are found in undercooked meat or poultry. They cause the disease toxoplasmosis. In pregnant women, toxoplasmosis can cause miscarriage or health problems for the baby. Following food safety guidelines and properly cooking meat can help prevent foodborne diseases from bacteria and parasites.

## Food for Thought

1. Besides limiting your intake of processed meats, what are some other ways you can help your body stay healthy?
2. How does knowing about the health effects of processed meat affect how you think about your diet?
3. What do you think about practicing Meatless Mondays, or skipping eating meat weekly on Mondays?

Try filling your plate with naturally colorful foods, healthy fats, and lots of fruits and vegetables. Often, the more natural colors your plate has, the healthier it is!

# Chapter Four

## Preserving Your Health

Reducing the amount of manufactured meats in your meals doesn't have to be a loss. Instead, it can be a way to discover new foods and ways of cooking them. Not all healthy diets are the same, but they do have some elements in common. First, they load up on fruits and vegetables, which deliver many of the vitamins, minerals, and other nutrients you need. Next, they add a good dose of whole grains to provide fiber and energy. Finally, they include healthy animal proteins from sources such as chicken, turkey, fish, or dairy. Vegetarians (who don't eat meat) and vegans (who don't eat any animal products) meet their protein needs with plant-based options like beans, legumes, and soy. There are so many delicious food choices beyond hot dogs, sausages, and lunch meat. These nutritious options are not only good for you, but they're also better for the planet.

### Healthy Plates

A helpful tool for designing nutritious meals is the Healthy Eating Plate from the Harvard Chan School of Public Health.

The Healthy Eating Plate divides up an imaginary dish to show the ideal portions of food for each meal. Fruits and vegetables should make up one-half of the plate. Whole grains should fill one-fourth. The remaining one-fourth should be healthy proteins. Healthy plant oils—olive, canola, soy, corn, sunflower, and peanut—are an important source of the good fats your brain and body need. They can be added carefully to the plate as dressings or used in cooking.

The healthiest types of grains are whole grains, which are given this name because they contain the entire grain kernel. Examples of whole grains include whole wheat, oats, barley, quinoa, and brown rice. Whole grains are usually eaten in the form of bread, pasta, or cooked grains.

Dairy foods include milk, cheese, and yogurt. The Healthy Eating Plate guidelines recommend having one or two servings of dairy a day as part of your healthy proteins. Other healthy proteins are fish, poultry like chicken or turkey, beans, and nuts. Healthy proteins do not include processed meats, and it is suggested that red meat is eaten only rarely.

Most of the fruit and vegetable section of the plate should be filled with vegetables. Vegetables can be raw or cooked. They cover all the colors of the rainbow and deliver a variety of vitamins, minerals, and nutrients. Carrots, green peas, broccoli, spinach, sweet potatoes, eggplant, green beans, squash, and onions are all good choices. Potatoes, on the other hand, should not count toward the vegetable portion. Everyone should eat 2 to 3 cups of vegetables each day. Add 1 ½ to 2 cups of fruits, such as apples, oranges, bananas, grapes, melon, and berries, to complete the colorful mix.

## Manufactured Meat Alternatives

Avoiding all processed meats may not be possible for some people. Fortunately, everyone can think carefully about the

# A Healthy Plate

vegetables and fruits   whole grains   protein

> Instead of measuring your foods, the Healthy Eating Plate can help you proportion the fruits and vegetables, whole grains, and proteins on your plate.

kinds of processed meats to eat and how frequently to eat them. If you do have processed meats, try to balance them with healthy whole grains, fruits, and vegetables.

Instead of using packaged lunch meats, you can try making a sandwich with fresh vegetables and hummus on whole-grain bread.

When you have processed meat for a special occasion, look for hot dogs and sausages that are low fat and low sodium. Turkey and chicken hot dogs may be lower in fat than beef or pork, but not always. Healthier lunch meat choices also exist. Chicken, turkey, or roast beef tend to be lower in fat and salt than ham, bologna, or salami.

Another option is to skip the processed meats entirely and eat a piece of lean beef, turkey, chicken, or fish, all of which are better sources of protein. The body needs protein to grow and for cells to develop.

Vegetarian or vegan hot dogs, sausages, and veggie burgers tend to be healthier than meat or poultry because they are made with soy, vegetable proteins, or grains like quinoa. These products are naturally low in fat and cholesterol and high in protein. Keep an eye out for options that are low in sodium too.

Some people may need to avoid certain processed meats altogether. People with a family history of cancer, diabetes, and heart disease may choose to be extra careful about the meats they eat. In addition, pregnant women, the elderly, children, and people with weakened immune systems must be cautious about bacteria and parasites from products like lunch meats.

## Food Safety

Manufactured meats are less likely to cause illness if they are prepared and stored safely. Most hot dogs are already cooked, but they should still be heated until they are steaming to prevent listeriosis and other infections caused by bacteria. People shouldn't leave hot dogs—even cooked ones—sitting out of the refrigerator for more than 2 hours (no more than 1 hour in warm temperatures). To avoid getting sick, it's important to throw out any unused hot dogs within a week of

# Lab-Grown Meat

The United Nations (UN) has suggested that lab-grown, or cultured, meat can be a beneficial option for people who are avoiding traditional meats for health and environmental reasons. Lab-grown meat is a new development in the food industry. Rather than raising animals on factory farms, meat is grown in a laboratory from animal **stem cells**. The cells multiply quickly in a controlled environment and can be ready to eat within weeks. This technique has been used to produce chicken, beef, pork, and fish.

Lab-grown meat has been promoted as an alternative to factory farm practices that are cruel to animals. It has also been supported by the UN and other international organizations as a way to reduce the high greenhouse gas emissions from animal agriculture and devote less natural area to farmland. On the other hand, some people are unsure about eating this food. It's a challenge to grow cells into an appetizing product with all the tastes, smells, and textures of traditional meat. Some people also worry about the safety and ethics of engineering meat. Lab-grown food can be designed to have healthier fats and nutrients, but even cultured red meat should be eaten less frequently than other meats.

---

opening the package. Uncured hot dogs may need to be tossed even sooner.

Heat any uncooked beef, lamb, pork, or veal until a cooking thermometer stuck in the middle reads 160 degrees Fahrenheit (71 degrees Celsius). Ground turkey or chicken should be cooked to 165°F (74°C). After handling raw meats, people should always wash their hands in soap and water. They also

Lab-grown meat factories may release less methane into the environment than cattle farms do, but they still use fossil fuels. Over time, carbon dioxide emissions from the artificial meat industry could be even more damaging to the environment than traditional practices, according to an Oxford University study.

need to thoroughly clean any cooking surfaces that the raw meat touched.

When buying canned meats, look at the date on the can to make sure the food has not expired. Do not buy cans that are dented, cracked, or bulging. It's also wise to clean the top of the can before opening it.

When cooking meat at home, it's always a good idea to use a thermometer and be sure the temperature is high enough to kill any bacteria that could make you sick.

## Moderation

The key to eating healthy is to enjoy most foods in moderation, or without going to extremes. Not everyone will choose to avoid processed meats entirely, but they can still be careful not to eat them too often. Eating a healthy diet of fruits, vegetables, whole grains, dairy, and lean meats can help balance the occasional hot dog or bologna sandwich.

A healthy lifestyle depends on more than just nutritious foods. Moving your body is another ingredient. According to government guidelines, people should aim for at least 30 minutes of walking, dancing, bicycling, weight training, swimming, aerobics, or other movement every day. Finding an activity that you enjoy doing will help you feel motivated to move.

Finally, it's important to be an informed eater. When people read package labels, they know exactly what ingredients they're about to eat. They also know how much fat, salt, and nutrients are in their food. Being an educated eater can help you live a long and healthy life.

## Food for Thought

1. Eating healthy is a lifelong practice. What choices can you make today for a healthier future?
2. Would you eat lab-grown meat? Why or why not?
3. Why is moderation important?

# Glossary

**attention-deficit/hyperactivity disorder:** A complex and common mental health condition that causes people to struggle with impulse control, focusing, and organization.

**botulism:** A type of food poisoning caused by *Clostridium botulinum* bacteria.

**byproduct:** Meat that comes from other parts of the animal, including the heart, kidneys, or liver.

**chronic:** Always present, as in a long-lasting disease or condition.

**contaminate:** To soil or infect by contact.

**curing:** A process of smoking and adding salt and other preservatives to meat to help it stay fresh longer.

**fermenting:** A process of breaking down carbohydrates in organic substances with microorganisms to preserve food.

**filler:** Nonmeat ingredients that are added to sausages to give them more substance and weight.

**high blood pressure:** An increase in blood volume in the arteries, which can force the heart to work harder.

**hyperactive:** A state in which a person is overly excited and has difficulty remaining calm.

**nitrosamine:** A cancer-causing substance that can form in the body when it is exposed to nitrites and nitrates.

**obesity:** A condition in which too much fat is accumulated and stored in the body.

**preservative:** A chemical added to food products to help them last longer without spoiling.

**ruminant:** A plant-eating, hoofed mammal with a complex stomach.

**stem cell:** A simple cell that can become a cell with a special function.

**stroke:** A blockage that can cause damage by preventing blood from flowing to an area of the brain.

# Find Out More

## Books

Evans, Kim Masters. *Farm Animals*. Farmington Hills, MI: Gale, 2018.

Martin, Claudia. *Vegetarian Food*. New York, NY: Enslow Publishing, 2019.

Perritano, John. *Food Safety*. Broomall, PA: Mason Crest, 2018.

## Websites

**Humane League**
*thehumaneleague.org/article/what-is-processed-meat*
The Humane League provides an argument for avoiding processed meat in order to develop a healthier lifestyle and oppose cruelty to animals in factory farms.

**My Plate**
*www.myplate.gov/*
The USDA offers a website with guidance on how to balance the foods you eat that's based on their publication, *The Dietary Guidelines for Americans*.

**World Health Organization**
*www.who.int/news-room/fact-sheets/detail/healthy-diet*
The World Health Organization provides guidance for people around the world on how to create a healthy, well-balanced diet.

# Organizations

**American Cancer Society**
3380 Chastain Meadows Pkwy NW, Suite 200
Kennesaw, GA 30144
*www.cancer.org/*
The American Cancer Society is committed to researching and preventing cancer while also providing services and support for people diagnosed with cancer.

**American Diabetes Association**
2451 Crystal Drive, Suite 900
Arlington, VA 22202
*www.diabetes.org/*
The American Diabetes Association aims to prevent diabetes through education, support people diagnosed with diabetes, and work to find a cure.

**American Heart Association**
7272 Greenville Avenue
Dallas, TX 75231
*www.heart.org/*
The American Heart Association raises awareness about heart disease, teaches people the signs of heart attack and stroke, and urges prevention of heart disease through healthy lifestyle choices.

**Publisher's note to educators and parents:** Our editors have carefully reviewed these websites to ensure that they are suitable for students. Many websites change frequently, however, and we cannot guarantee that a site's future contents will continue to meet our high standards of quality and educational value. Be advised that students should be closely supervised whenever they access the internet.

# Index

## A
advanced meat recovery (AMR), 8
American Heart Association, 10
arterial disease, 16
artifical coloring/dyes, 5, 7, 11–12

## B
bacon, 4, 25
bacteria, 13–14, 29–31, 37, 40
Bisphenol A (BPA), 6

## C
cancer, 5, 11, 20–21, 23, 26, 28, 37
canned meats, 4, 6, 39
chemicals, 5, 7, 12
cholesterol, 15–18, 21, 24, 37
climate change, 20
colon, 21, 26
cooking, 26, 31, 33–34, 38–41

## D
diabetes, 5, 20, 23–24, 26, 37

## E
environment, 20, 38–39

## F
fish, 4, 15, 25, 33–34, 37–38

## G
greenhouse gases, 20, 38

## H
ham, 4–6, 12, 37
Healthy Eating Plate, 33–34
hot dogs, 4–5, 7–13, 15, 17–18, 25, 27, 30, 33, 37–38, 41

## I
immune system, 31, 37
insulin, 24
intestines, 9, 26, 29–30

## K
kidneys, 18, 24, 26, 30

## L
lab-grown meat, 38–39, 41
listeriosis, 31

lunch meat, 4–5, 12, 19, 28, 33, 36–37
lung disease, 23, 28–29

## M
mad cow disease, 8
mechanical separation, 8
movement, 25, 41

## N
nitrates, 11–13, 18, 21–22, 26, 28
nitrites, 12–13, 18, 21–22, 26

## P
parasites, 29, 31, 37
portions, 34
protein, 7, 9, 15, 18, 33–34, 37

## R
red meat, 5, 17, 20, 25, 34, 38

## S
salt, 5, 10
saturated fat, 9, 17
sausage, 4–5, 7, 9–10, 25, 41
sodium, 10, 12, 15, 18–19, 37

## T
temperature, 15–16, 31, 40
thermometer, 38, 40

## U
unsaturated fats, 9, 15–16, 21

## V
vegan, 33, 37
vegetarian, 33, 37
vitamins, 7, 9, 15–16, 21

## W
weight, 8, 15, 23–24, 41
whole grains, 24–25, 33–34, 37, 41
World Health Organization (WHO), 28

# About the Author

**Rachael Morlock** is a freelance writer and copyeditor. She is the author of several nonfiction and picture books for children and enjoys researching new subjects through her work. In the past, she worked as the coordinator of a Community Supported Bakery. Rachael is always interested in supporting local farmers and food producers in Western New York, where she lives with her dog, Rilke.